DINO-BIRDS

From dinosaurs to birds

Angela Milner

THE NATURAL HISTORY MUSEUM

Introduction

What do *Tyrannosaurus rex* and *Erithacus rubecula* (the common European robin) have in common? Much more than you might realise; the robin in your garden is a modern dinosaur, albeit a small one, but perhaps just as aggressive as its huge and distant relative. How could it be possible that such a small, delicate bird could have descended from such a lumbering meat-eating giant? Birds and dinosaurs seem so different; sufficiently different that birds have traditionally been classified separately as feathered and (mostly) flying animals.

So are birds really the descendants of meat-eating dinosaurs? Yes, indeed they are. Spectacular feathered fossils, discovered in Liaoning Province, north east China since 1996 have provided the final clues in a long detective story that began 140 years ago. They also fulfil predictions made by many palaeontologists; if birds are the descendants of meat-eating dinosaurs (technically called theropods), then those dinosaurs must have had feathers too. They did, and the Chinese fossils prove it beyond all doubt.

Tyrannosaurus, *a giant meat-eater*
that lived 65 million years ago…
… was a distant relative of this
small garden dinosaur, a robin.

Ancient wing

Debate about the origin of birds goes back to the discovery of the first specimen of the earliest known bird, *Archaeopteryx*, in the Solnhofen limestone in Bavaria in 1861. This is one of the principal treasures in The Natural History Museum, London and one of the world's most famous fossils. It consists of most of the skeleton of a magpie-sized animal split between two slabs of limestone, like two facing pages of a stone book. But more than that, the exceptionally fine-grained rock bears clear and bold imprints of beautifully detailed feathers arranged round both forelimbs and along both sides of the tail.

This *Archaeopteryx* tombstone was formed from fine plastic mud on the bottom of a calm shallow lagoon 147 million years ago, at the end of the Jurassic Period. The Solnhofen limestone in Bavaria is famous for the preservation of many fossils. These sediments are extremely fine grained and compact, and have preserved the imprints of *Archaeopteryx* flight feathers – feathers that are exactly like those of modern flying birds. The feathers belong to an animal that had toothed jaws, three fingers ending in sharply pointed claws and a long bony tail; hallmarks of theropod dinosaurs.

Archaeopteryx *with clawed fingers, toothed jaws and wings, was part bird and part dinosaur.*

The skeleton clearly shows other characteristics shared with birds such as a reversed perching toe on the hind foot and a wishbone, properly called a furcula, formed from the fusion of the two collar bones or clavicles, that braced the middle of the chest. The *Archaeopteryx* wishbone has a simple boomerang shape compared to the long, springy, V-shaped wishbone in a chicken.

It seems, therefore, that *Archaeopteryx* represents a snapshot of evolution 'caught in the act'; a dinosaur equipped with wings and capable of flight, an ancient bird about the size of a modern magpie.

Small theropod-type teeth lined the upper jaw of Archaeopteryx.

The main slab of the Archaeopteryx *specimen at The Natural History Museum, London, preserves most of the skeleton and impressions of the wing and tail feathers.*

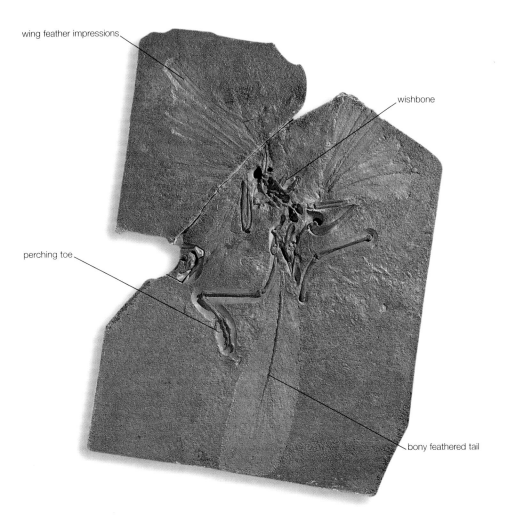

wing feather impressions

wishbone

perching toe

bony feathered tail

Archaeopteryx *had the same number and arrangement of primary (long outer) and secondary (shorter inner) flight feathers as modern birds – in this case a magpie.*

Archaeopteryx has always been regarded as an important fossil. Richard Owen, the first director of The Natural History Museum, London, went to great lengths to purchase it from Germany in 1862. The specimen was named *Archaeopteryx lithographica* by Hermann von Meyer, a leading German palaeontologist in 1861, and was collected that year by quarry workers extracting the Solnhofen limestone for lithographic printing plates. The specimen was owned by Dr Carl Haberlein, a local Bavarian doctor. Carl Haberlein amassed a huge collection of Solnhofen fossils, including *Archaeopteryx*, by accepting them as payment in kind from the quarry workers for medical services. He put his collection of nearly 2000 fossils up for sale in 1862 to raise a dowry for his daughter's wedding. Richard Owen paid £750 (US$1000) to secure the whole collection for The Natural History Museum, London. He wasted no time in describing *Archaeopteryx* in detail in 1863 as a primitive, long-tailed bird and simply left it at that.

It was Thomas Henry Huxley, Charles Darwin's great supporter, who first suggested the dinosaur connection. He noted many similarities between the hind limb bones of *Megalosaurus*, one of the few meat-eating dinosaurs then known, and those of the ostrich. He also recognised a suite of bony features common to *Archaeopteryx* and a small theropod dinosaur called *Compsognathus*, discovered in the Solnhofen limestone in 1859. Huxley's arguments were not sustained due to lack of evidence and the fact that not much was known about dinosaurs in the 1860s. Today *Compsognathus* is recognised as belonging to a group called coelurosaurs. The name is often used to describe small, lightly-built theropods. It also defines a diverse group of theropod dinosaurs (including *Tyrannosaurus*) all of which share features that indicate they are closer to birds than meat-eaters such as *Allosaurus*. Among these features are elongation of the bony processes that link the neck vertebrae together, and a short ischium (the backward pointing lower hip bone).

A second specimen of *Archaeopteryx*, even more complete, came into the hands of Dr Haberlein's son, Ernst, in 1877. Accounts of how much he paid for the specimen vary from 140 to 2000 deutsch-marks. Ernst Haberlein offered it to the Bavarian State Collections for more than ten times the latter price, which they could not meet. After extensive preparation work which revealed just how complete a specimen it was, Haberlein reputedly offered it, plus the rest of his collection, to Yale University for 36,000 marks, a sum which the great Professor O.C. Marsh at Yale refused to pay. Eventually the German industrialist Werner von Siemens advanced the sum of 20,000 marks to keep it in Germany where it is now the pride of the Berlin Natural History Museum. The Berlin specimen shows clearly the dinosaurian toothed jaws and the long clawed hands. The composition of the wings, divided into primary and secondary flight feathers, is comparable with any bird today as is the asymmetrical structure of the individual feathers, with unequal length barbs (the structures making up the vane of a feather) on each side of the shaft. It appears that *Archaeopteryx* had modern wings attached to a small carnivorous dinosaur. It must have been able to fly, but how well has been the subject of much debate.

The spread-out wings and long clawed fingers show clearly on the Berlin Archaeopteryx.

A further five skeletons of *Archaeopteryx* have turned up from the Solnhofen limestone quarries in the past 130 years, the most recent one in 1993. Not all are complete but they all show some evidence of feather impressions. Interestingly, the identity of several skeletons as being *Archaeopteryx* were not recognised at first. A fragment collected in 1858 showing feather impressions, claws and a knee joint had been identified as a flying reptile, a pterosaur, and was exhibited as such in the Teyler Museum in Holland. John Ostrom, an American palaeontologist, noticed the feather impressions that gave away its true identity while visiting the museum gallery in 1970. Two other specimens with less clear feather impressions were mistaken initially for the little coelurosaur *Compsognathus*. This alone tells us something that Huxley had already noticed!

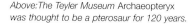

Above:The Teyler Museum Archaeopteryx *was thought to be a pterosaur for 120 years.*

Left: The short forelimbs and long hind legs of Compsognathus *indicate that it was a fast runner.*

Right: Compsognathus, *a 40 centimetre long coelurosaur from Solnhofen.*

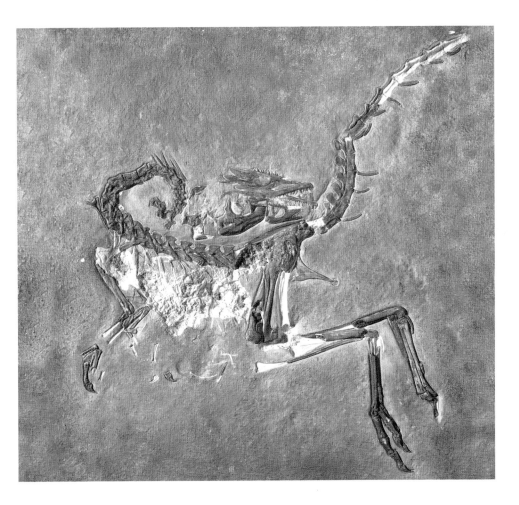

A Danish artist and illustrator, Gerhard Heilmann, was the next to consider the origin of birds in a book written in 1916 and translated into English in 1926. He too observed that small meat-eating dinosaurs, including *Compsognathus*, shared many features with *Archaeopteryx*. They would be ideal ancestors for birds except that one important link was missing. *Compsognathus* apparently had no clavicles at all; neither fused like *Archaeopteryx* nor separate. Heilmann concluded that dinosaurs could not have given rise to birds since they had lost this vital character. He could not envisage that lost bones might be regained during evolution, so he reasoned that birds must have descended from much more archaic fossil reptiles that did possess clavicles. He proposed a hypothetical ancestral form called 'Proavis'. From that point on, the whole problem of bird origins stagnated until 1969 when John Ostrom of Yale University described an astonishing new dinosaur that he called *Deinonychus*.

Gerhard Heilmann envisaged the hypothetical ancestor of birds as a tree-climbing, gliding animal covered in feathers.

Enter the 'raptors'

Deinonychus, *a slender and agile predator was equipped with long, mobile grasping hands.*

Deinonychus is a 3 metre (10 feet) long, lightly built, fast-running predatory dinosaur from early Cretaceous rocks (110 million years old) in Montana, USA. It belongs to a family of advanced theropods called dromaeosaurs, or the 'raptors', which also includes *Velociraptor* from the Gobi Desert in Mongolia. Dromaeosaurs had a swivelling, sickle-shaped, slashing claw on each back foot used for attacking their prey. *Deinonychus*, and probably the other dromaeosaurs too, were pack hunters, using their slashing claws, as well as grasping hands tipped with sharply curved claws and jaws full of ripping teeth, to tackle prey much larger than themselves. Modern equivalents are hyaenas or African hunting dogs that hunt cooperatively to bring down large herbivores such as antelope and buffalo.

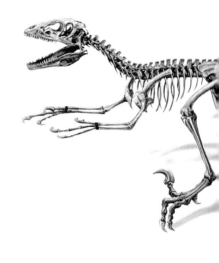

Dromaeosaurs were as intelligent as some birds today, and their lifestyle suggests that they must have had a high and constant body temperature to maintain it. John Ostrom noticed many skeletal features that dromaeosaurs shared with *Archaeopteryx*, including long forearms and very long three-fingered hands ending in curved claws. The key feature was the presence of a half-moon shaped bone in the wrist; a 'semi-lunate carpal'. This allowed the wrist to be flexed sideways in addition to up and down movements. Dromaeosaurs could fold their long hands almost in the manner that birds do today. Another advantage was that the hands could be swivelled and rapidly whipped forwards to grab prey. It is no coincidence that this

The overall skeleton design of Deinonychus (upper) and Archaeopteryx (lower) is very similar. Both had swivelling wrist bones (marked in red) that allowed a wide range of sweeping and folding hand movements.

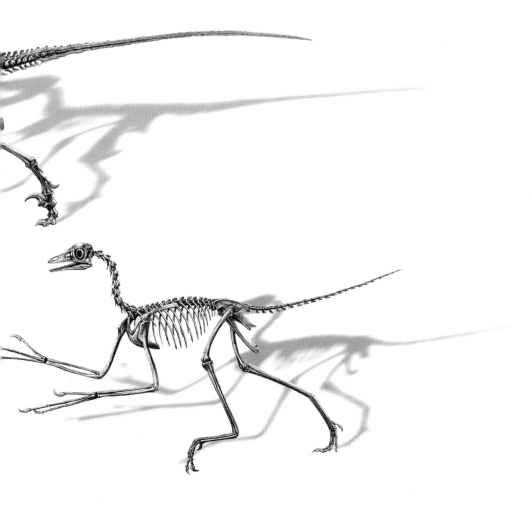

movement is similar to the flight stroke of a bird. This ability to rotate the wrists was shared with other small, advanced theropods such as oviraptorids and troodonts. All three families of long-handed theropods are collectively called maniraptorans.

John Ostrom's work in the 1970s rekindled not only the debate about the dinosaur origin of birds, but also whether dinosaurs were hot blooded. This debate has been going on ever since. Almost all palaeontologists have long been convinced that Huxley, and more recently, John Ostrom, were right about bird ancestry. The only missing piece of evidence was a theropod with feathers or wishbones. Now we have both. Theropod dinosaurs do have wishbones, and some of the earliest still had separate clavicles. A theropod with a wishbone was actually known as far back as 1924, but the structure was misidentified as a V-shaped belly rib (all meat-eating dinosaurs have a corset of belly ribs to support the abdomen). Modern studies have demonstrated the presence of a wishbone in a wide range of theropods related to birds. Heilmann's objection to theropods as bird ancestors was thus overcome; it was just unfortunate that the furcula was missing in *Compsognathus*.

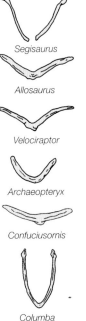

Segisaurus

Allosaurus

Velociraptor

Archaeopteryx

Confuciusornis

Columba
(pigeon)

Fascinating discoveries about dinosaur eggs and nests have brought even more evidence to support the case. Fossils of dinosaurs can tell us much more than how they are all related to each other. Dinosaurs laid hard-shelled eggs; that has been known for more than a century. What is more, many dinosaurs laid their eggs in carefully constructed nests in which the eggs were neatly spaced and carefully positioned. Some were colonial nesters, using the same breeding sites each season. In 1993 an American Museum of Natural History expedition to Mongolia discovered an example of dinosaur parental care in action. An *Oviraptor*, a small 1.6 metre (5.2 feet) long theropod with a bizarre shaped head and toothless beak, was discovered crouching on top of its nest of eggs.

Wishbones began as paired collar bones in early theropods and became fused together to form a V-shaped 'furcula' in modern birds.

This parent was sitting on its nest, legs folded and arms outstretched around its clutch of eggs when it was overwhelmed suddenly by a slumping sand dune. This provides an example of dinosaur brooding behaviour, frozen in time; the dinosaur parent was doing 75 million years ago exactly what most modern birds do today.

Here was dramatic evidence that birds have inherited the nesting habits of their dinosaur ancestors, habits that probably go even further back in time than the *Oviraptor* discovery. There is, as yet, no evidence of how *Tyrannosaurus* cared for its eggs, but fossil

Caught in the act, a parent Oviraptor *fossilized while sitting on its nest, forelimbs curling protectively round its eggs.*

A nesting Oviraptor *reconstructed. The orange area indicates the bones found in the nest.*

nests of some large plant-eating duck-billed dinosaurs or 'hadrosaurs' demonstrate an alternative way of incubating eggs – covering them with sediment and vegetation to control the temperature. Crocodiles also do this by building nest mounds and some birds called megapodes do it too. This demonstrates just one of the links between birds and crocodiles. Birds and crocodiles share many anatomical features that are not found in any other living animals. On that basis, the two are more closely related to each other than either is to any other living animal group. The fossil record of dinosaurs shows all the evolutionary changes that took place between crocodiles and birds.

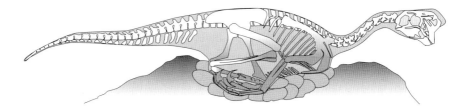

Feathers from China

Beginning in 1996, Chinese palaeontologists announced a series of remarkable and exciting dinosaur discoveries. These confirmed that nearly all the features in which birds differ from all other animals were also present in dinosaurs long before birds evolved. The first of these finds was a small, juvenile theropod dinosaur, very like *Compsognathus*, but with a fine filamentous 'downy' covering, which the Chinese named *Sinosauropteryx*. It came from a locality called Sihetun near the town of Beipiao in Liaoning Province, north east China. Here was the first evidence of a theropod dinosaur with a feather-like coat, composed of simple hollow filaments running from nose to tail. It caused a sensation. Here at last was what palaeontologists had been waiting for; the predictions about feathered dinosaurs had come true and the last piece could be fitted into the jigsaw puzzle of bird origins.

Fossil hunting in Liaoning. Huge quarries have been dug to reveal the very fine fossil-bearing layers, which then have to be split carefully to reveal their treasures.

LIAONING

Fossil sites.....

Beijing ●

CHINA

Examples of the exquisite Liaoning fossil preservation: a slab with a fish, mayfly larvae and a lizard (left); a mammal complete with its furry coat (below) and a long-necked freshwater reptile (right).

Liaoning localities are unique in that actual soft tissues are preserved. The fossils are preserved in fine-grained sediments that preserve exquisite details. Around 120 million years ago the area was a forested lakeside environment, subject to frequent volcanic activity. The flora and fauna were periodically overwhelmed with volcanic gas and ash, burying and preserving them almost perfectly. Millions of plants, insects, molluscs, crustaceans, fish, frogs, salamanders, turtles, lizards and mammals were suffocated and ended up on the lake bed. They provide a vivid picture of the rich variety of life in that corner of Asia in early Cretaceous times.

Sinosauropteryx was a small two-legged predator. It had toothed jaws equipped with flattened serrated teeth, the pattern typical for meat-eating dinosaurs. Its forelimbs were short and ended in clawed fingers. The hind limbs, especially long from the knee downwards, are typical of fast runners. Besides the juvenile Sinosauropteryx which is 55 centimetres (22 inches) long from nose to tail, adult skeletons tell us that it grew to at least 150 centimetres (59 inches) long. One of the adult skeletons also shows us what it had eaten for its last meal, the lower jaw bones of a mammal are still in position near the end of the gut. Sinosauropteryx probably snapped up anything small enough to catch, lizards as well as mammals.

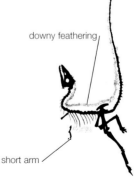

downy feathering

short arm

The dark structures running from head to tail along the back of this juvenile Sinosauropteryx *are a mat of simple hollow 'downy' filaments.*

Sinosauropteryx *restored –
a small, fast running
predator that has just
caught a lizard. The simple,
hollow, feather-like coat
would have provided
insulation to help keep the
body temperature constant.*

This adult Sinosauropteryx *was over a metre long. The remains of its last meal are still inside the gut – a pair of mammal jaws are lodged in between the pelvic bones.*

Right: The variety of feathered dinosaurs from Liaoning indicate that feathers go back a long way in theropod history.

mammal jaws

The outer covering of *Sinosauropteryx* does not look like the branched feathers of modern day birds. It seems to consist of single, hollow, feather-like fibres that may represent an early stage in the origin and development of feathers. A second Liaoning coelurosaur, the therizinosaur *Beipiaosaurus*, shows a similar covering of even longer filamentous structures.

These discoveries enable predictions to be made about how far back down the family tree of dinosaurs feather-like coverings appeared. If coelurosaurs had feathers, then so did all the groups that evolved after them, including tyrannosaurs, oviraptorids and dromaeosaurs.

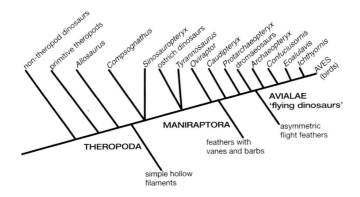

a

b

c

d

e

Modern bird feathers consist of a quill embedded in the skin that emerges as a shaft from which barbs sprout. Flight feathers have a strong rigid shaft, and barbules interlock adjacent barbs together to form a continuous structure. The barbs are asymmetrically arranged along either side of the shaft to give an aerofoil shape with leading and trailing surfaces to produce lift. Down feathers that cover the body are symmetrical with a slender shaft and do not have the aerodynamic qualities needed for flight. Contour feathers define the body outline, and have symmetrical vanes and varying proportions of downy bases. Richard Prum, at the University of Kansas, has proposed a model of how feathers may have originated. Feather follicles first gave rise to simple hollow sheaths (1) that gradually became more complex over time to produce increasingly elaborate feathers, down-like ones first (2,3) and then, by different pathways, feathers with parallel barbs (4–7) and barbules and eventually asymmetric vanes (8).

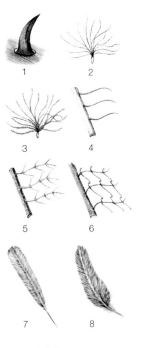

A modelled sequence of feather evolution.

Left: Modern bird feathers: (a) down; (b,c) flight; (d,e) contour.

So why did feathers arise? Insulation is the most likely explanation. A coat of hollow filaments would be useful in helping to keep heat in, especially in small-bodied and young animals. The simple physical ratio of surface area to volume means that small bodies have a larger surface area relative to their volume, and so loose (or gain) heat more rapidly than larger bodies. It might seem very far-fetched to envisage a feathery *Tyrannosaurus*. A 6 tonne beast had a great deal of bulk to buffer temperature changes, but its chicks (as yet unknown) could have been as fluffy as the farmyard version to keep them warm while they grew quickly.

Feathered dinosaur discoveries from Liaoning have come thick and fast since the initial discoveries. They show in fantastic detail how feathers became elaborated and adapted for different functions.

Protarchaeopteryx is known from a single, frustratingly incomplete individual first described in 1997. It was clearly a long-legged, running theropod with no suggestion of a backward pointing toe.

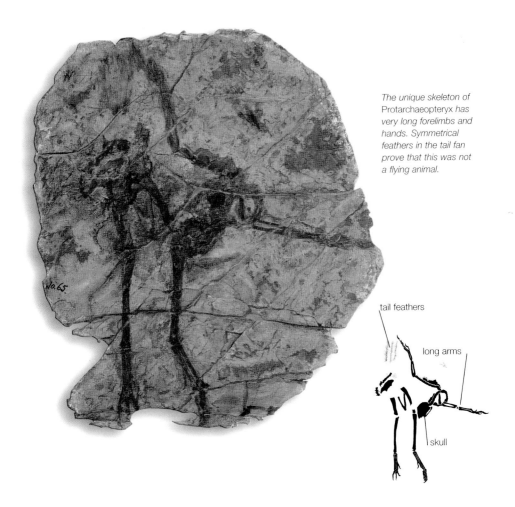

The unique skeleton of Protarchaeopteryx has very long forelimbs and hands. Symmetrical feathers in the tail fan prove that this was not a flying animal.

No.65.

tail feathers

long arms

skull

It had swivelling wrists for seizing prey, very long arms and huge, sharply clawed hands, almost as long as those of *Archaeopteryx*. It had a short tail with a clump of feather at the tip. These are true feathers with a central shaft and symmetrically arranged barbs. A few scattered feathers near one of the forelimbs show the same structure. They are not flight feathers like the asymmetrical ones possessed by *Archaeopteryx*.

Proarchaeopteryx was therefore not a flier, those long arms with swivelling wrists had evolved first for catching prey. The tailfeathers might have been used for some other function – display perhaps?

Caudipteryx, also named in 1997, provides a dramatic example of feathers for show, as it had a jaunty fan of feathers on the end of its tail. Several individuals (probably adults), all about the size of a turkey, have been found in the Sihetun area. *Caudipteryx* was a very long-legged runner with powerful hind limbs and unusually short arms tipped with claws. It also had quite a short skull with just a few small teeth at the tip of the upper jaws and a beak.

Protarchaeopteryx, *a very long-legged and long armed predator chasing down a lizard.*

A heap of stomach stones (gastroliths) had been preserved in most individuals, a gastric mill in life that functioned to grind up hard food, aided by the muscular walls of the gizzard (the powerful grinding stomach). Farmyard chickens pick up grit to do the same job. So what kind of lifestyle can we deduce for *Caudipteryx*? It appears to have been a fast runner with a diet that included hard food gathered with a beak; it could have eaten fruits and seeds from the lush vegetation around the lakeside, or fed on shelled crustaceans and molluscs that lived in the lake shallows.

Left: Caudipteryx' *hand is fringed with long, perfectly preserved symmetrical feathers.*

Right: The Caudipteryx' *small, almost toothless skull and gastric mill suggest that it fed on small hard food. This specimen is preserved on a bed of shelled crustaceans that could have been part of its diet.*

small skull

short arms

gastroliths

tail fan

Caudipteryx *might have used its tail fan for display like these black grouse cocks, 'lekking' to attract a mate.*

Caudipteryx has beautifully preserved long symmetrical feathers attached to its very short forelimbs. These arms were impossibly short for any thoughts of flying so what were the feathers for?

Perhaps they could have been used for display in conjunction with the tail. *Caudipteryx* is thought by some palaeontologists to be related to the beaked oviraptorids, such as the Mongolian *Oviraptor*. Its arms may well have been feathered too, but they would not have been preserved in the coarse desert sands.

Sinornithosaurus, found in 1998 confirmed that the family closest to *Archaeopteryx* were feather covered; just what we would expect to find. Perhaps the most stunning dino-bird of them all was a juvenile dromaeosaur discovered in 2000. It was covered from head to tail in fine branched feathers and is affectionately known as 'Dave the fuzzy raptor'.

Right: 'Fuzzy raptor' was covered in branched feathers.

Far right: The spectacular preservation of 'fuzzy raptor', a juvenile dromaeosaur, shows the typical long grasping hands and swivelling wrist bones, as well as the extensive feather preservation.

generate enough downward force to climb near vertical surfaces. Dial calls this behaviour 'wing assisted vertical running'. Perhaps this escape mechanism could have been the starting point for the development of flight; the fluffy forelimbs of small dinosaurs could have been just as effective as those of game bird chicks.

Microraptor is the smallest dromaeosaur known. Its very curved hand and foot claws leave no doubt that it was a capable tree-climber.

Feathers and flight paths

Liaoning's feathered dinosaurs led to the inescapable conclusion that dinosaurs did indeed possess feathers and that they evolved initially for purposes other than flight. It may be that a simple insulating cover arose first and was later modified for display, signalling and finally flight. How did dinosaurs get off the ground and learn to fly?

Two rival theories as to how dinosaurs became airborne have emerged since John Ostrom's landmark study of *Deinonychus*. Each has vigorous support across the scientific community. The 'ground up' theory proposes that long-legged, ground-dwelling maniraptorans ran fast using their sweeping arm action to help gain enough lift to take off into the air. The 'trees down' theory holds that small dinosaurs were climbers and launched themselves from a height, using their 'proto-wings' to glide from tree to tree or from tree to ground.

John Ostrom, in 1974, was the first to suggest that getting airborne was a by-product of running predators using their sweeping arm action as insect traps. The folding, swivelling, prey-

Trees down or ground up? Flight might have started arboreally from feathered tree-climbing gliders or cursorially from fast ground runners.

snatching arm action was already present and could have helped to generate lift by moving the arms in a wing-like motion with the hands pivoted forwards. Alternatively, running dinosaurs might have held their arms out sideways to balance. Graham Taylor of Oxford University suggested, in 2002, that feathers, even simple ones that lacked modern feather construction, could have provided drag to help steer during a predatory leap. Dromaeosaurs such as *Sinornithosaurus* could have fitted this model, taking off almost by accident. However, *Sinornithosaurus* grew to at least 150 centimetres (59 inches) long and weighed 20–30 kilograms (44–66 pounds). *Deinonychus* probably weighed 60–75 kilograms (132–165 pounds). Such animals would have had to run very fast indeed to get airborne and to generate enough lift to overcome the drag of the body weight and long tail. As soon as they left the ground they would lose any thrust effect and tend to stall. Experts on aerodynamics are very sceptical that flight developed in this way.

Could *Archaeopteryx* with its modern flight feathers and a body weight of less than 0.5 kilograms (1 pound) have taken off from a running start? It was certainly capable of active flight, but lacked the anatomical improvements and manoeuvrability of modern birds for taking off, landing, steering and flying slowly. The maximum running speed of the Berlin *Archaeopteryx* is about 2 metres (6.5 feet) per second, and the stalling speed is about 6 metres (19.5 feet) per second, so it could have launched from a perch well above the ground or run into the wind to get airborne. Engineering calculations suggest that the wing area and muscle mass of *Archaeopteryx*, compared to its bulk, was probably insufficient to provide it with enough power for a ground-based start. Relying on a windy day for a quick escape from a hungry predator would not have been a very dependable strategy. The alternative, of running and climbing out of reach, might seem a better survival option.

Archaeopteryx was certainly at home on the ground but it was equipped for climbing too. Its claws on both hands and feet are multipurpose in shape; suitable for walking and curved enough for clinging and climbing. The horny claw sheaths are preserved on the toes of some *Archaeopteryx* specimens. They are sharply pointed and lack any signs of worn or blunted tips, so perhaps their owners did not pound along the ground, wearing them down.

The smallest dromaeosaur from Liaoning, *Microraptor*, provides a test of the arboreal habits of these animals. With a trunk length of just 47 millimetres (1.9 inches), *Microraptor* is the smallest adult non-avian dinosaur known. It was feathered, as traces of long branched feathers are preserved attached to the forelimbs and legs. Its toes, as well as its fingers, were equipped with slender and very curved claws; the degree of claw curvature falls within the range of modern climbers. The Chinese scientists who described *Microraptor* in 2000 suggest that it was a tree dweller not a ground runner; arboreal rather than cursorial.

Research in progress by Dr Ken Dial, of Montana State University, has some interesting implications about how flight could have originated. He has shown that game birds, such as partridges and pheasants that spend most of their time on the ground, use their wings to help them climb a tree or bush when danger threatens. They beat their wings vigorously as they climb. This action provides a downward force that helps to hold them against the surface, rather like the skirts on a racing car that hold it down on the road. Even chicks with stubby, downy wings are able to

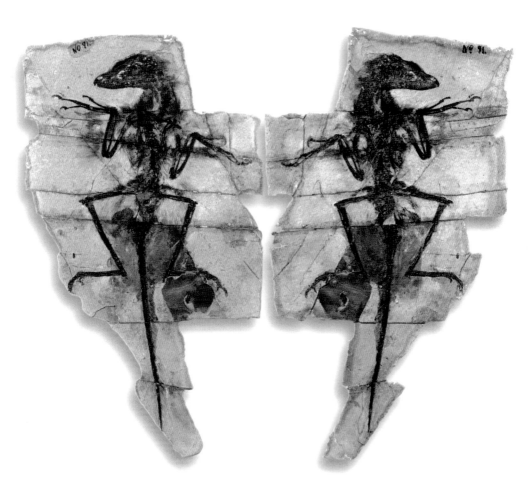

Early birds

Many fossils of flying birds have been found in the Liaoning deposits. They allow a glimpse of how the basic flight kit was modified and improved to make flight more efficient. *Confuciousornis* and *Changchengornis*, which lived between 124 and 122 million years ago, are among the oldest known birds after *Archaeopteryx*. They are also the oldest known beaked birds. Both had toothless jaws and their jaw bones show evidence that they were covered in a horny beak in life.

Confuciousornis, the commonest of the two, about the size of a rook, is represented by hundreds of individuals. The sheer numbers preserved suggest it lived in large colonies around forested lake margins. *Confuciousornis* was more bird-like than *Archaeopteryx* with modern flight apparatus, although it retained fully functional grasping claws on all but the index finger that supported the flight feathers. The wings were very large with exceptionally long flight feathers, longer than the body, a one-off adaptation to increasing lift indicating that it had diverged from the direct line leading to more advanced birds. Its body was compact and streamlined with an enlarged breastbone to anchor flight muscles and strut-like bones

*Liaoning birds may have lived in large
social colonies like these rooks and jackdaws.*

bracing the shoulder to the chest. The long, cumbersome, bony tail was much reduced, well on the way to forming a pygostyle, the tight knot of bones that form the tail stump (the parson's nose) and control the tail feathers in modern birds. Sexual dimorphism is clearly visible in *Confuciusornis*; males are slightly larger than females and have a pair of exceptionally long tail feathers.

Left: Confuciusornis *was sexually dimorphic. Males had a pair of long tail pennants to advertise for females.*

Right: The long wings of Confuciusornis *are clearly preserved. This male shows the two tail pennants extending from an elongate blob of bone – the pygostyle – at the end of the short tail.*

This is the only specimen known of starling-sized Changchengornis. *It had a horn-covered beak and a well-developed perching foot.*

Right: Changchengornis *had long tail feathers like* Confuciusornis, *but it is not known whether both sexes possessed them.*

Changchengornis, of which there is only a single specimen described, also had long tail feathers. The foot had a longer perching toe and the claws are more curved than *Confuciousornis,* suggesting it may have had a greater grasping ability. Both had the ability to fly and take off from the ground.

Liaoxiornis, also from the Liaoning deposits is about 2 million years younger than *Changchengornis* and *Confuciousornis.* It is the smallest early Cretaceous bird known. Not much bigger than a wren, it provides a glimpse of the rich variety of birds that existed in the lakeside forests of Liaoning.

Liaoxiornis, *with its head turned to the right, was about the size of a wren.*

Early Cretaceous birds have been found in other parts of the world as well as China, many as small as sparrows, and these also chart the course of flight improvements. The Spanish birds, *Iberomesornis* and *Concornis* show further elongation of the forelimbs, shortened tails, and expansion and elaboration of the shoulder and chest bones to support bigger muscles for a more powerful flight stroke. *Eoalulavis* sported the first known alula, a separate bunch of flight feathers attached to the thumb. The alula can be raised and lowered to control flight at low speeds, in much the same way that the flaps on an aeroplane wing prevent stalling at low speed when an aircraft approaches landing. An alula is also known in several Liaoning birds.

Above: Archaeopteryx' *second finger supported the primary feathers, this was flattened and modified in later birds and an alula developed on the 'thumb'.*

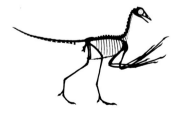

All of these modifications allowed low speed flight and increased manoeuvrability. Modern wing design had been perfected by around 125 million years ago. It preceded the changes to the hips and hind limbs that we see in modern birds. A swinging thigh bone provided the walking stride in theropods. In modern birds, a knee flexion mechanism moves the legs backward and forwards; birds essentially walk from the knees down. This change in pattern is linked with the reduction in the tail, the loss of tail-hip musculature and the consequent moving forwards of the centre of gravity that sits around the middle of the trunk in modern birds. This extensive remodelling of the back half of birds' skeletons and muscles took another 50 million years or so; fully modern birds began to appear by about 70 million years ago in late Cretaceous times.

Dinosaurs' body shape changed radically as they took to the air; they lost their long bony tails and moved their centre of gravity forwards.

Conclusion

The Liaoning fauna has provided us with a snapshot of an ancient community in which different groups of feathered dinosaurs and flying birds were all living alongside each other. How were they all living together if birds descended from dinosaurs? A further complication is that *Archaeopteryx* is 27 million years older than the feathered theropods from Liaoning. The key point to remember is that Liaoning reveals what was living at that time in that part of the world, not when the various groups evolved. The Liaoning animals are representatives of several lines of evolution, some of which had existed for longer than others.

The Chinese dino-birds have given us many clues that suggest the ways in which dinosaurs could have taken flight, but many stages along the way to airworthiness are still hidden in the rocks of the Earth's past history. There is still no evidence of exactly how asymmetric flight feathers developed from downy feathers, how long ago it happened and how the number and arrangement of flight feathers in the wing was established. We need some Liaoning-type preservations much earlier in time, before the appearance of *Archaeopteryx* in the latest Jurassic, to provide the answers.

How to pronounce

Archaeopteryx	Ar-kee-OP-ter-iks	= ancient wing
Cathyornis	Kath-AY-or-niss	= China bird
Caudipteryx	Caw-DIP-ter-iks	= tail feather
Changchengornis	Chang-cheng-OR-nis	= Chinese bird
Compsognathus	Komp-SOG-nay-thus	= pretty jaw
Confuciusornis	Con-few-shus-OR-nis	= Confucius bird
Deinonychus	Dine-ON-i-kus	= terrible claw
Dromaeosaurid	Drom-Ay-ee-oh-sore-rid	= (member of) running lizard family
Liaoxiornis delicatus	Lao-shee-OR-nis	= Western Liaoning bird
Microraptor	Mike-ROW-rap-tor	= little plunderer
Oviraptor	Oh-VEE-rap-tor	= egg thief
Protarchaeopteryx	Pro-tark-ee-OP-ter-iks	= first ancient wing
Sinornithosaurus	Sine-ORN-ith-oh-sore-us	= Chinese bird-lizard
Sinosauropteryx	Sino-sor-OP-ter-iks	= Chinese dragon wing
Tyrannosaurus	Tie-RAN-oh-sore-us	= tyrant lizard
Velociraptor	Vel-OSS-ee-rap-tor	= quick plunderer

63

Author's acknowledgements

I am very grateful to the Director of the Geological Museum of China, Cheng Liwei for permission to include pictures of NGMC specimens in this book. I also thank Cheng Liwei and Jia Zhongpeng for their generous hospitality during my visits to Beijing. It is a pleasure to record my thanks to Phil Crabb, Frank Greenaway and Tim Parmenter of The Natural History Museum's Photographic Unit for their skill and support in providing most of the photographs. Lastly, Lorraine Cornish, Phil Crabb, Jane Mainwaring and Gail Nolan provided companionship and support during visits to China.

First published by The Natural History, Museum, Cromwell Road, London SW7 5BD

© The Natural History Museum, London 2002

ISBN 0-565-09174-3

Edited by: Rebecca Harman
Designed by: Dina Koulama
Reproduction and printing by: Craft Print, Singapore
Front cover: 'Fuzzy raptor', a juvenile dromaeosaur